CHOLESTEROL

Written by

Dr. David G. Williams

ISBN 0-944649-03-3

Copyright 1988
Mountain Home Publishing

All rights reserved. No part of this book may be reproduced or utilized in any form or by any means, electronic or mechanical, including: photocopying, recording or by any information storage and retrieval system, without permission in writing from the Publisher. Inquiries should be addressed to Mountain Home Publishing, P. O. Box 829, Ingram, TX 78025.

CONTENTS

INTRODUCTION..................................1

CHAPTER 1
Cholesterol--You Can Run From It, But You Can't Hide..................................3

CHAPTER 2
A Little Respect Please..................................7

CHAPTER 3
Big Jobs For Little Tubes..................................9

CHAPTER 4
Can Cholesterol Be House Trained?
(Is Cholesterol Really the Problem?)..................................11

    Where There's Conflict, There's Opportunity..................................17
    Plastic Fat..................................18
    Another $Billion Market..................................19
    Summary..................................21

CHAPTER 5
How Much Is Too Much Of A Good Thing?........23

CHAPTER 6

The Truth About Eggs, Butter and Fat.................29

    Eggs..29
    Oil, Fats, Butter & Margarine.......................32
    Poly-unsaturated Fats..................................35
    Hydrogenation..36
    Mono-unsaturated oils.................................40
    Saturated Fats..41
    Butter VS. Margarine..................................42
    Meats..43
    Fish...43
    Pork..44
    Red Meat...44

CHAPTER 7

Natural Proven Methods To Lower Cholesterol..47

    Various Types of Fiber................................49

CHAPTER 8

Some Not-So-Natural, But Interesting Ways To Lower Cholesterol..61

    GH3..61
    Plasma Pharesis...62

CHAPTER 9
Cholesterol Testing Tips..65

CHAPTER 10
Saving Someone's Life..69

CHAPTER 11
Saving Your Own Life..71

CONCLUSION..73

The approaches described in this book are not offered as cures, prescriptions, diagnosis's or a means of diagnosis to different conditions. The information must be viewed as an objective compilation of existing data and research without the author's, or publisher's endorsement. The author and publisher assume no responsibility in the correct or incorrect use of this information, and no attempt should be made to use this information as a form of treatment without the approval and guidance of your doctor.

INTRODUCTION

One of the most controversial and confusing health subjects today is cholesterol. For decades it has been at the center of many heated debates. For years it received top billing in hundreds of medical and health journals. But none of this can even come close to the recent surge of publicity focusing on cholesterol. Lately it has become "the" health topic of every magazine, book, radio and television show in the country. The American Medical Association and the federal government have practically declared war on cholesterol. The American Heart Association and the National Heart, Lung and Blood Institute have branded it "the number one cause of heart attack". Just when you've finally been convinced it's the worst thing you could put in your body another article surfaces showing that the cholesterol from your diet has very little if anything to do with your blood levels. As you read further the situation becomes even more confusing.

One source claims you should never allow an egg to rest on your breakfast table again. Someone else claims eggs are special, and come equipped with ingredients to neutralize their own cholesterol. Well if you can eat eggs again, what about butter instead of margarine? Milk? Before it gets too confusing and you have to consider eliminating 90% of your whole diet you might want to see if you even need to lower your cholesterol levels. After all, you may be one of the lucky ones who's levels are already low enough. Finding out exactly what is considered safe and unsafe isn't that easy either. Sure the new guidelines have just been released, but each time they set new ones they're lower than the ones before. Many say these latest ones aren't low enough and they don't give enough consideration to the two types of cholesterol. After all, H.D.L. cholesterol is better than L.D.L. cholesterol. Isn't it? And if all of this isn't enough to make you want to forget about cholesterol altogether, you learn cholesterol isn't a fat. But fats do influence cholesterol levels and of course there would have to be good fats and bad fats just to make things even more complicated.

Before heading for the woods to live on wild plants and tree bark, or running for the nearest fried chicken franchise in desperation, take a few minutes to read through this book and clear up the confusion.

In just a short while you'll be able to confuse everyone else with your new found knowledge about cholesterol, but more importantly you'll be able to separate the facts from fiction and know exactly what steps you can take to lessen your chances of dying from heart disease. (You'll also learn you <u>can</u> have a little butter on your toast with your breakfast eggs!)

CHAPTER 1

Cholesterol--You Can Run From It, But You Can't Hide

You might call me a slow learner (behind my back please). When it comes to learning about any new and unfamiliar topic I like to start at the beginning--the very beginning. Learning about cholesterol is no different. Before you declare war on anything (if that's what you decide to do) you should know exactly what it is and what it does. After finishing the first paragraph of most articles on cholesterol you can't help but visualize chunks of yellow fat pumping through your bloodstream. You end up with as much respect for cholesterol as you have for a cockroach in your kitchen. You and I both know cholesterol has to have some good attributes (the cockroach ... I'm not so sure -- maybe it's holding the cure for cancer or something). Seriously we couldn't survive without cholesterol. Let's take a closer look at exactly what this stuff is all about.

Cholesterol is a modified fat called a sterol. It's actually more like a wax than a fat or an oil. It doesn't dissolve readily in water or your blood stream either, since blood is mostly water. The majority of cholesterol in your body is manufactured by the liver and smaller amounts by the small intestine and individual cells throughout the body. Between 1,500 mg. and 1,800 mg. are produced by the body everyday. Only between 200 mg. and 800 mg. daily come from the average American diet.

If you totally eliminated all cholesterol from your diet your body would compensate by slightly increasing production if it needed additional amounts. Increasing cholesterol in your diet on the other hand would cause liver production to cut back in able to maintain consistent levels. It has been shown time and time again that dietary cholesterol has only minimal influence on blood cholesterol levels. A 1978 study showed that blood cholesterol levels did not reflect dietary intake of between 200 mg. and 900 mg. of cholesterol a day (Slater and Alfin-Slater, "Effect of Dietary Cholesterol on Plasma Cholesterol, H.D.L. Cholesterol, and Triglycerides in Human Subjects," Fed. Proc. Abstr. 2194, 1 Mar. 1978). In the famous Farmington study 437 men and 475 women were observed for up to ten years and there was no correlation between the dietary intake of cholesterol and serum cholesterol levels (Gordon, et al, "High Density Lipoprotein as a Protective Factor Against Coronary Heart Disease: The Farmington Study," Amer. J. Med. 62:707-708 May 1977).

By now it should be obvious that you couldn't get rid of all of the cholesterol in your body, even if you wanted to, and eliminating it from your diet isn't the way to lower blood levels. After learning about all the good things cholesterol does, it may make sense why our cholesterol production is so constant. The way cholesterol problems have been handled thus far we should all be thankful for the divine intervention that keeps cholesterol production out of our control.

CHAPTER 2

A Little Respect Please

Cholesterol performs a lot of essential functions in the human body that lately it hasn't received much credit for. In fact, it is so important that every cell in your body manufactures and contains cholesterol.

Every cell uses cholesterol to help construct its protective cell membrane. This waxy substance helps the cells maintain their shape and at the same time forms a special barrier against various substances trying to enter or exit the cell. On a larger scale it provides an additional protective barrier in the skin.

Cholesterol in the skin helps make it resistant to the penetration of certain liquids and keeps our most important liquid from leaving our body too quickly. Without cholesterol it is estimated that you would lose 4 to 5 gallons of water per day through evaporation instead, of the usual 10 to 14 ounces. You can thank cholesterol for not having to drag a garden hose around with you or stand next to a water fountain all day.

The basic framework of cholesterol also provides the basis of several very important steroid hormones. These include those produced in the adrenal glands, the ovaries and the testes. It also plays an important role in helping make vitamin D.

The majority of cholesterol (about 80%) is used by the liver to help produce bile salts. Bile salts are stored in the gallbladder and used to help in the digestion and absorption of fats in the diet.

If you hadn't watched television, listened to radio or read any printed material for the last fifty years and had just read all the great things I've written about cholesterol, you might like it. You might even want to add more to your diet. But you know better, since high cholesterol levels have been associated with increased chances of heart disease. That brings up the next question we need to answer. Just what does it do to the heart and blood vessels? It's really a little more involved than to say it just clogs up our arteries. If we look a little closer at what happens to a blood vessel we might see that cholesterol isn't totally at fault.

CHAPTER 3

Big Jobs For Little Tubes

Most of us would pick the heart as the single most important organ necessary to keep us alive. After all, once it stops beating, death soon follows. Second on the list of importance we could list our blood vessels. They transport the necessary oxygen and nutrients to keep our other organs alive. When someone mentions circulation, we automatically think about the heart and sometimes forget the over 60,000 miles of blood vessels we each have.

Blood vessels are not just simple tubes that carry blood from one part of the body to another. They are complex living organs that consume oxygen, burn energy and excrete waste. They have nerves, layers of muscles and even <u>their own small blood vessels</u> to supply their different layers with oxygen and nutrients. They are subjected to constant pounding pressure and friction from passing blood. Contracting and expanding in response to signals from nerves, hormones, drugs, chemicals and even nutrients from food and vitamins are only a couple of their duties. They also absorb nutrients and excrete waste material directly to and from the blood. Their inner walls must be constantly repaired with strong new cells as old ones become worn and fragile. Your blood vessels never rest; for like the heart, to rest would mean certain death.

As the cells lining arteries die and sluff off, they leave a small indention or roughened area. This newly exposed area is sensitive and susceptible to inflammation. As blood rushes past areas of inflammation, it has a tendency to deposit certain fats and cholesterol. A gradual buildup in these areas can lead to a narrowing or blockage. When the blood flow is restricted to organs and tissue beyond the blockage, they die from lack of oxygen. Depending on the location, this could result in a stroke or heart attack.

CHAPTER 4

Can Cholesterol Be House Trained? (Is Cholesterol Really the Problem?)

As you've probably gathered, cholesterol is essential for life and it really only becomes a problem when it starts to accumulate in your arteries. It's sort of like a new pet, most of the problem is eliminated once it's house broken and doesn't create a mess. To put it another way, if cholesterol does so many good things for us and there's no way we can totally get rid of it, maybe we need to concentrate more on trying to keep it from causing any harm. Granted, too much of anything can be harmful and lowering your cholesterol levels can without a doubt make a difference. Later I'll show you all kinds of natural ways to do just that. But first, let's take a closer look at the question, "Is cholesterol really the problem?".

This may sound like a stupid question considering all the hoopla and media announcements condemning cholesterol, but, just so we won't be accused of being closed minded, let's consider a few other ideas.

First, what we're really talking about is atherosclerosis or clogging or hardening of the arteries. Nobody would really care about cholesterol if it weren't involved in this process. "While mental illness is our greatest socio-economic crippler, cancer our greatest enigma, arthritis and rheumatism our greatest crippler and accidents our greatest disgrace, atherosclerosis is by far our greatest killer", R. L. Holman (1958). When you take into account, heart attacks, kidney failure and all the problems directly linked to atherosclerosis it is no exaggeration to say that <u>it causes one-half of all deaths.</u> To put it another way, we're talking about over 750,000 deaths a year. It's like having ten jet airliners, each carrying over 200 people, crashing every single day of the year!

Atherosclerosis kills. It kills by cutting off the life sustaining blood supply to vital organs like the heart, brain, kidneys etc. Our real concern should be what causes an artery to clog. What makes cholesterol and fat accumulate in certain arteries, while leaving others untouched? Why can some people eat large amounts of fat and not be bothered with atherosclerosis or high cholesterol levels? Many have automatically labeled cholesterol the <u>cause</u> of atherosclerosis simply because it is found in the clogs or plaque deposits. If you cut your hand and it became infected, a trip to the doctor for antibiotics might leave you with the impression that germs and bacteria were the source of your problem. Germs weren't the problem-- carelessness resulting in a cut was the problem. You can't eliminate all the germs and bacteria in the world just as you can't eliminate all cholesterol. Cholesterol has been <u>linked</u> to atherosclerosis for years and so has age, saturated fats, alcohol, insulin, smoking etc. etc. Why then, has everybody lately declared cholesterol "the major cause of coronary-artery disease"? Has new scientific evidence and medical studies finally discovered that cholesterol is the undisputed villain causing damage to our heart and arteries? Have they found that cholesterol is the single most important substance that leads to atherosclerosis? Hardly.

Here's a few of the things we know about atherosclerosis. It is dependent on many factors and not any one particular agent.

    A. Fats, especially saturated animal fats, can increase atherosclerosis. These harmful fats raise blood cholesterol levels. They easily become rancid and oxidized giving off free radicals, which can inflame arterial walls. We also know that some unsaturated fats may retard atherosclerosis. Some vegetable and fish oils lower cholesterol levels and act as antioxidants to neutralize harmful free radicals.

    B. Saturated fats tend to make blood cells stick together. This causes blood flow in the arteries to slow down, leading to clumping and sticking to artery walls. The clumps later become sites of calcium and cholesterol deposition--atherosclerosis.

    C. Blood vessels as I mentioned earlier, absorb nutrients. They also act as filters. Once the stomach and intestines have digested and broken down fat, carbohydrates and proteins, they are dumped into the bloodstream. Some particles are carried to different organs for storage or for immediate use. Other particles are absorbed and filtered through artery walls. Particles, like fats, that can't dissolve in water may not pass completely through the blood vessel wall. These "foreign" particles may accumulate in a particular area forming a plaque or blockage.

D. The violent pounding, from high blood pressure and/or a weakness in the tiny blood vessels that supply the different layers in artery walls, can lead to a rupture or hemorrhage. When these tiny vessels within an artery break, fat circulating in the blood is released and tends to accumulate in these areas.

E. Certain areas of blood vessels are subjected to enormous amounts of stress. One area in particular, is the very small arteries that supply the heart muscle. These tiny arteries branch <u>directly</u> from the aorta and unfortunately have no way to step-down the high pressure blood flow they receive. It is these same coronary arteries whose blockage leads to heart attack.

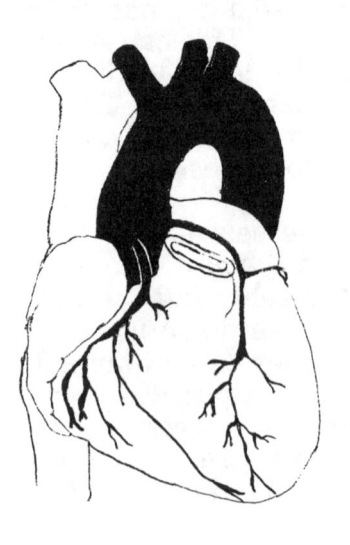

The small coronary arteries that supply the heart muscle, branch directly from the high pressure aorta.

Where arteries branch, fork or bend sharply, additional stress and friction can damage the structure of the vessel walls and result in fat and calcium deposition.

F. Age is known to play a role in atherosclerosis. Like other tissues, our arteries lose their ability to stretch and rebound to their original shapes and sizes. With time, blood vessels can become brittle and crack. Calcium, cholesterol and other particles traveling in the blood, become embedded in these small cracks and fissures. With accumulation and time these areas begin to form the plaques of atherosclerosis.

G. There is no doubt, heredity plays a role in atherosclerosis. A past family history of heart and blood vessel disease makes you more susceptible to the same type of problems. Many times, whole families eventually die from the same cause, like stroke or heart attack. On the other hand, some people seem to inherit arteries so sturdy that none of the factors we've mentioned seem to effect them.

Heredity can also influence cholesterol and fat levels in the blood, blood pressure, digestion etc., all of which play important roles in atherosclerosis.

Everyone should examine their family's health history. Knowing and compensating for certain inherited weaknesses may avoid unnecessary illness and an early death.

H. Hormones play a vital role in the development of atherosclerosis. This alone explains why the problem is so very common in men, while women seem almost immune to the problem, <u>up until but not beyond menopause.</u> The female hormone, estrogen, seems to be the protecting agent. One particular research study showed that atherosclerosis could be prevented in male chickens being fed a high cholesterol diet, if estrogen was added to the food. (No one to my knowledge has ever advocated the use of estrogen in humans for preventing atherosclerosis. Quite obviously, the side-effects of such treatments would be worse than the disease.)

You can probably guess by the large number of factors that have been linked to clogging of the arteries, that there is still a considerable amount of confusion surrounding the subject. I should also mention that in the last century, nearly everyone of these factors, at one time or another, has been promoted as the sole cause of atherosclerosis.

Getting back to the most popular concept today, cholesterol, no one has shown for sure that cholesterol is the cause of atherosclerosis. Most everyone still agrees that it only plays a role in the process.

Cholesterol is a popular scapegoat today for a simple reason. It is generally very easy to lower blood cholesterol levels---even without medication, as you'll see later. We can't yet change our age, our inherited weaknesses or the arrangement of our circulatory system, but we can change our diet.

We eat way to much animal fat, fried foods, lard, margarine, butter, grease, etc. In less developed countries, where the diet consists of nuts, fruits, grains and vegetables, heart and artery disease are almost unheard of. (These places are getting harder to find as we gradually introduce the "civilized ways" of life to the far corners of the planet. You would think we were on a mission from God, considering the time and effort we spend trying to bring everyone a neighborhood hamburger or fried chicken franchise.)

Nutritional pioneers, like Nathan Pritikin, have demonstrated how a strict low fat diet can lower cholesterol and blood fat levels, which in turn seem to stop and possibly even reverse atherosclerosis and heart disease. Strict dieting can make a difference. The main objection to this type of program is the unwillingness to give up our "civilized" foods. How many times have you heard someone say, "I'm going to eat some of that-(ice cream, pie, candy,etc.) even if it kills me", or "Life just isn't worth living if I can't have a little -(fried fish, hamburger meat, etc.) every once in a while."?

Where There's Conflict There's Opportunity

There's one thing you can't deny about civilized man (and woman)--they're creative. We're always looking for a way to do things easier. We definitely want to have our cake and eat it too, and this doesn't go unnoticed. The financial riches in discovering ways that allow us to indulge in our habits and not have to pay the price with our health, are staggering. Food manufacturers and drug companies make billions of dollars trying to cater to this want.

PLASTIC FAT

Proctor & Gamble has filed a petition with the Food and Drug Administration (FDA) to approve its new discovery--a fat substitute. The actual name of the product is Olestra, a form of sucrose polyester (SPE). Like many "great" discoveries, it was an accident. P & G scientists were actually trying to formulate a product that would provide quick nourishment for premature infants. Instead, they chemically bound table sugar (sucrose) to fatty acids and created SPE, which the body can hardly absorb at all. I call it plastic fat.

Supposedly, Olestra can be substituted for cooking oils or shortening and used for frying or to make ice cream, pies, cakes, etc. It looks, tastes and feels like fat, but it can't be broken down in the digestive tract since we don't have the digestive enzymes for this man-made product. It can't be absorbed, so there's no calories. Adding to all this, is that Olestra is supposed to lower harmful cholesterol levels by attaching and pulling it out of the body. (Many predict P&G will apply later to use it as a drug, if it gets approval as a food additive.)

In theory, P&G's new plastic fat will allow you to eat all of the fatty foods you want without paying the price with clogged arteries and heart disease. Just think, all the fried foods, pastries, ice cream you want--it sounds too good to be true. When you check deeper into it, it may just be that.

Serious questions have been raised about the safety of Olestra. Some of P&G's own research raises questions about Olestra and increased pituitary tumors, leukemia and liver disease in laboratory animals. It is also thought that SPE mixes with fat soluble vitamins (A,D,E,F,K) and interferes with their absorption. Hopefully, additional long-term studies will be required before plastic fat starts showing up in our food.

Don't expect P&G to give up on bringing Olestra to the market. It seems the public is willing to pay a pretty steep price to have their fat. Estimates are the current market for Olestra is over $1 billion a year.

Another $Billion Market

Drug companies have a knack for supplying the wants of the American public. The majority of the American public wants the easy way out when it comes to their health. They spend hundreds of millions of dollars each year on antacids, head and chest cold remedies, laxatives and appetite suppressants in an effort to "treat" problems directly related to improper diet and lifestyle. Show the drug companies a "need" and you can be sure they'll be happy to supply the answer. What might surprise you, is their ability to even create a new need through huge advertising and promotional campaigns. As a prime example of this let's look at the recent interest in cholesterol.

For years there have been several cholesterol-lowering drugs on the market including: cholestyramine sold as Questran by Bristal-Myer Co., Colestid from Upjohn Co. etc., but they either had undesirable side-effects or an unpleasant taste. A couple of new anti-cholesterol drugs just happened to be approved by the FDA and released for sale to coincide with the newly declared war on cholesterol. The timing is simply uncanny. Does it seem even the least bit strange, that just as federal government agencies, medical associations and health organizations announce their programs to lower deadly cholesterol, a couple of drug manufacturers have announced their release of some anti-cholesterol "miracle" drugs?

Most public health campaigns come on the heels of some newly released medical research which has discovered the so-called cause of some disease. In contrast, it appears the recent anti-cholesterol campaign was announced to coincide with the release of some new pharmaceutical products. The publicity provided by this "unusual coincidence" has boosted the anti-cholesterol drug market to an estimated $1 billion a year. Projected new sales of Merck & Co.'s new anti-cholesterol wonder drug, Lovastatin, have caused the company's stock prices to triple in the last two years. (A one year's supply of Lovastatin costs around $2,600.) Warner-Lambert's new anti-cholesterol drug, Gemfibrozil, is expected to boost the company's sales from $60 million to $200 million a year.

Undoubtedly, other new anti-cholesterol drugs will be hitting the lucrative market as more doctors and patients become convinced cholesterol is the root of all evil. While the publicity campaigns concentrate on increased drug sales, some doctors are worried about the long term use of these unproven drugs. "There is not nearly enough experience to judge serious long-term side-effects," says Dr. Virgil Brown, a leading blood specialist and president of Washington D.C.'s Medlantic Research Foundation. Plenty of promising drugs have been heavily hyped, only to fall by the wayside once side-effects turned up. (Forbes, Nov.2,1987)

SUMMARY

No one has shown cholesterol to be the cause of atherosclerosis. It certainly has been associated with heart and artery disease, just as a host of other factors have. Cholesterol is found in the plaques that develop and eventually clog arteries, but there has to be some damage or weakness in the artery walls before cholesterol, calcium and blood fats begin to attach and form plaques.

Highly elevated cholesterol levels are symptoms of deeper problems, usually of a dietary nature. Diets lacking sufficient fiber, certain vitamins and minerals, and high in saturated fats, cause high blood cholesterol levels. These same factors make it harder for cholesterol to stay in solution or dissolve in the blood. Cholesterol not in solution has a tendency to "settle out" and accumulate on arterial walls, whereas even high blood levels of cholesterol in solution can pass throughout the body without causing problems.

Cholesterol is being touted as the cause of atherosclerosis and heart disease, subsequently new drugs have been released to lower cholesterol levels. Since cholesterol is only a symptom of deeper problems, it should be obvious that these drugs can never be the total answer. Many authorities not only question their use, but also their long-term safety.

CHAPTER 5

How Much Is Too Much Of A Good Thing?

Arriving at the ideal blood cholesterol levels would seem like an easy task, but various experts have continued to change what are considered ideal levels. To further complicate matters, there are several different types of cholesterol. Let's briefly look at a couple of different types.

Cholesterol, if you remember, doesn't dissolve well in blood and because of this, it would have a hard time traveling through the bloodstream by itself. To make things easier, it hitches a ride on fatty proteins. The most important fatty proteins are H.D.L. (high-density lipoproteins) and L.D.L. (low-density lipoproteins). The type of lipoprotein cholesterol attaches to, determines if it is H.D.L. or L.D.L. cholesterol.

H.D.L. hangs on very tightly to the cholesterol it carries. It won't let it get loose to attach to arterial walls. In some cases, it may even snatch up additional cholesterol already stuck to a wall reducing the size of a plaque or build-up.

H.D.L. keeps cholesterol in solution and moves it safely throughout the body. For these reasons, H.D.L. cholesterol is considered to be good cholesterol.

L.D.L. cholesterol on the other hand, is considered to be bad cholesterol. L.D.L. deposits its cholesterol on the walls of arteries. Instead of reducing the size of plaques, L.D.L. adds to them.

(Another type of bad cholesterol is called beta V.L.D.L., which stands for beta-very-low-density lipoprotein. This type deposits cholesterol in the small arteries supplying the heart. It receives very little attention, since it makes up only a small portion of the total blood cholesterol.)

Obviously, from what I've said, you'd want higher levels of H.D.L. cholesterol and lower levels of L.D.L. cholesterol to stay healthy. Determining just how high or how low isn't that easy. (If they made all of this medical stuff too easy to figure, half a million doctors would be out of work applying for your job.) In a recent attempt to simplify matters, the National Institutes of Health (NIH) released a new set of cholesterol guidelines.

The new guidelines basically list three cholesterol levels that apply to everyone regardless of age.

NATIONAL INSTITUTE OF HEALTH CHOLESTEROL GUIDELINES:

Total Cholesterol (L.D.L. & H.D.L.):
Desirable — Less than 200mg/dl
Borderline High — From 200-239mg/dl
High — Equal to or above 240 mg/dl

These guidelines don't even take into account the difference between good (H.D.L.) and bad (L.D.L.) cholesterol unless you have a total cholesterol of over 240 mg/dl. or two or more of the following risk factors:

1. Being male
2. Family history of coronary heart disease (CHD)
3. Cigarette smoker
4. High blood pressure
5. An H.D.L. cholesterol level below 35 mg/dl
6. Diabetes mellitus
7. History of vascular disease
8. Severe obesity

If you have two or more of these or total cholesterol over 240mg/dl then you might want to figure your bad cholesterol or L.D.L. cholesterol levels.

L.D.L. cholesterol=
(total cholesterol) - (H.D.L. cholesterol) - (triglycerides)/5

The guidelines for L.D.L. cholesterol are as follows:

Desirone   Less than 130 mg/dl
Borderline High   From 130-139 mg/dl
High   Equal to or above 160 mg/dl

OTHER GUIDELINES

The NIH guidelines have been criticized by various groups for being too high. Others feel more emphasis should have been placed on the levels of beneficial H.D.L. levels.

Advocates of the strict Pritikin program, suggest that total cholesterol levels should be 100 + your age and should never exceed 160 mg/dl.

Those who tend to give more credit to the benefits of H.D.L. cholesterol have a formula to determine your heart attack risk.

$$\text{RISK FACTOR} = \frac{\text{Total cholesterol level}}{\text{H.D.L. cholesterol level}}$$

*A risk factor less than 4.5 is considered good and acceptable, from 7 to 9 doubles your risk of having a heart attack and above 13 triples the risk.

SUMMARY

No one knows the best cholesterol levels for optimal health. I would suspect they fall somewhere between the NIH levels and the ones recommended by Pritiken, but it must be kept in mind that your cholesterol level is just one of the many factors associated with heart disease. Smoking, obesity, high blood pressure and diet all contribute to heart attack and atherosclerosis.

Rather than totally concentrating on lowering cholesterol levels, let's cover some of the known causes of heart and artery disease. Not surprisingly, many of these causes also tend to raise harmful cholesterol levels. Eliminate the cause and you eliminate the symptoms, like high cholesterol.

CHAPTER 6

The Truth About Eggs, Butter And Fat

Most everyone agrees that a clean healthy diet should include more fruits, vegetables and whole grains and less saturated animal fats. However, there still remains a considerable amount of confusion and controversy surrounding foods like eggs, butter vs. margarine and various cooking oils. Let's take a look at each of these individually.

EGGS

For years the public has been told to avoid eggs, especially the high cholesterol yolk. The egg scare started in the 1950's and 1960's during an earlier promotional campaign condemning cholesterol. Patients and doctors alike still hold fast to the idea of no eggs, simply because eggs have a high cholesterol content. Many refuse to change their stance even when hundreds of studies have shown that the amount of cholesterol we eat has very little influence on our cholesterol blood levels.

If that weren't enough, specific studies have shown that consuming moderate amounts of eggs doesn't effect cholesterol levels.

A recent study reported in the British Medical Journal, Volume 294, Page 333, showed that 7 eggs a week combined with a low fat, high complex carbohydrate, high fiber diet did not raise cholesterol levels.

Egg yolks do contain a large amount of cholesterol and because of this, they've received a lot of bad publicity. What has been overlooked, is the fact that egg yolks are also one of the richest sources of choline, a component of lecithin. Choline acts like a fat and cholesterol dissolver. It keeps the cholesterol in the egg moving through the bloodstream and doesn't allow it to stack up on arterial walls. Eggs are also rich in minerals, vitamins and essential amino acids. For these and other reasons, it has been called one of the most perfect foods.

I have often thought that the egg industry should consider hiring the same people who started the egg-cholesterol-scare campaign. They must be some of the best people in the publicity business if after 20 or 30 years people are still frightened of eggs.

What about other high cholesterol foods-- should they be eliminated from the diet? Those who advocate the Pritikin diet say yes. Their program suggests eating no more than 100 mg. of cholesterol a day. The new NIH guidelines limit dietary cholesterol to 300 mg. a day. If you look at the following chart, you'll see why many people trying to follow these guidelines would be hesitant about including eggs in their diet.

CHOLESTEROL CONTENT OF SOME COMMON FOODS

Solids: 3 ounces   Liquids: 1 Cup

FOOD	Milligrams per 100 grams
Vegetables	0
Fruits	0
Nuts	0
Grains	0
Seeds	0
Soymilk	0
Cheddar cheese	27
Milk	40
Halibut	50
Haddock	60
Chicken (without skin)	60
Tuna	63
Beef	70
Pork	70
Lamb	70
Turkey (without skin)	82
Crab	100
Shrimp	150
Hamburger (6 oz.)	162
Lobster	200
Steak (8 oz.)	208
Liver	300
Egg	353
Beef Liver (6.oz.)	516
Cheese Pizza	600
Kidney	680
Chicken Liver (6 oz.)	1072
Brains	1700+

Looking at the chart, you'll notice that many of the high cholesterol foods are high in saturated fats(particularly the red meats). It's these harmful fats we have to worry about, not the cholesterol levels, since our intake of cholesterol doesn't really influence our blood cholesterol. Even though the seafood items have high cholesterol levels, many have been shown to actually <u>lower</u> blood cholesterol levels. In part, this is due to their Omega-3 oil content as we'll discuss shortly. Just because a food contains cholesterol, doesn't necessarily make it harmful. This is another instance where many have gone overboard on the cholesterol fear. Moderation is the key. Even though the study I mentioned showed that eggs added to an already clean diet didn't effect cholesterol levels, it would be foolish to eat eggs at every meal. Based on the previous chart, it would also be foolish to assume that beef was 5 times better for you than eggs because it only has 1/5 the amount of cholesterol.

Your whole health picture doesn't revolve around the cholesterol content of food. Granted it is worth the time and effort to lower your cholesterol levels <u>because the natural foods that lower cholesterol, (high fiber, fish oils, grains, etc.) do so only as an added bonus, while in the mean time, they improve your overall health and well-being.</u>

OIL, FATS, BUTTER & MARGARINE

Undoubtedly, this is one area where the food industry has used deceit and half-truths in an enormously successful effort to sell their products. Here again, to be able to protect yourself and family against atherosclerosis and heart disease, you need to have a clear understanding about the differences between cholesterol, saturated fat and unsaturated fat.

I'll start at the beginning and attempt to explain these topics as simply as possible. I may have to even teach you a little chemistry along the way. If it looks like it's getting too tough, take a break or just skip over this section of the book for the time being. (If you do decide to take a break, don't let me catch you heading for the corner chicken hut to order those heart-stopping chicken nuggets and french fries.)

First, let's talk about cholesterol again. A lot of oils and margarines claim to be cholesterol free. At first glance this might seem like an amazing feat, but in reality, all oils or fats that come from either plants or vegetables are naturally cholesterol free. Only animal products contain cholesterol, so all of those corn, soybean, palm and coconut oil products can say they are cholesterol-free. Animal products like, butter, milk, eggs, meat, seafood,etc., will contain cholesterol. Most companies producing oil and margarine are quick to point out their cholesterol-free products, but you know cholesterol isn't the real problem anyway, it's the type of fat (saturated or unsaturated) that we need to be concerned about.

There are two basic types of fat we need to talk about--saturated and unsaturated.

In chemistry, the word saturated means a substance has combined to the full extent of its combining capacity with another substance. In other words, a saturated fat is one that can't react or combine with other substances easily. Your body has a difficult time combining with or breaking down some saturated fats and as a result, they tend to accumulate in the body, clogging arteries and raising cholesterol levels.

In chemistry, everybody talks about bonds and bonding sites. This refers to the ability of one substance to combine with other substances. In the first illustration, you'll see a fat molecule that has open bonding sites. This makes it an unsaturated fat. In the second illustration, the saturated fat has no open bonding sites.

ILLUSTRATION #1 (Unsaturated)

```
    H H H H H H H H H H H H H H H H H OH
    | | | | | | | | | | | | | | | | | |
H - C-C-C-C-C-C=C-C-C=C-C-C-C-C-C-C-C-C-=O
    | | | | |   |     | | | | | | |
    H H H H H   H     H H H H H H H
```

ILLUSTRATION #2 (Saturated)

```
    H H H H H H H H H H H H H H H H H OH
    | | | | | | | | | | | | | | | | | |
H - C-C-C-C-C-C-C-C-C-C-C-C-C-C-C-C-C=O
    | | | | | | | | | | | | | | | |
    H H H H H H H H H H H H H H H H
```

Fats can be compared to a chain of molecules. If there are blank spaces in the chain where enzymes and nutrients can attach themselves, then the fat is unsaturated. (Poly=many. Poly-unsaturated fats have more blank spaces than mono-unsaturated fats.) Fats whose chains are full, so to speak, leave no room for other substances to attach--these are saturated fats.

Let's take a closer look at these two kinds of

fats--the unsaturated ones (poly and mono-unsaturates) and the saturated ones.

POLY-UNSATURATED FATS

Several years ago, researchers learned that poly-unsaturated oils could do a couple of things, like lower blood pressure and cholesterol levels. (More recently a new poly-unsaturated star called Omega-3 has gained in popularity. In addition to dropping blood pressure and cholesterol, it may help everything from allergies to arthritis.) You can probably recall the hundreds of television commercials and magazine ads promoting poly-unsaturated cooking oils and margarines. All of the advertising apparently worked. Since the beginning of this century, almost all of the fat increase in the American diet has come from unsaturated vegetable fats, rather than saturated animal fats. It's unfortunate that the advertisers neglected to mention two vital points. Slight reductions in dietary cholesterol do not lessen heart attacks and the increased consumption of poly-unsaturated fats may very well be contributing to the ever-increasing incidence of cancer.

Refined poly-unsaturated oils react easily with substances they come into contact with. The more unsaturated (the more chemical bonding sites available), the more reactive a substance is. While natural foods high in poly-unsaturates are protected by nature, refined oils have no such protection and can quickly become cancer-causing perioxidized fats.

Nuts and grains high in poly-unsaturates, contain numerous antioxidant nutrients as well as a compact protective fibrous capsule. Once the oil is removed from the nut or grain it becomes susceptible to rancidity when exposed to oxygen. This free-radical oxidative attack results in perioxidized fats that suppress the immune system, form blood clots, mutate cells and lead to cancer. Most nutritional authorities today, agree that the more poly-unsaturates you consume, the more antioxidant nutrients you need to remain healthy. Unfortunately, these are the very same nutrients that are being removed by modern food processing (vitamins A,B-1,B-5,B-6,C,E, the amino acid cysteine and the minerals selenium and zinc).

To make matters worse, the food industry has developed a process called hydrogenation.

HYDROGENATION

To increase the marketability of cholesterol-free poly-unsaturated oils, the food industry devised a method to convert liquid oils into semi-solid fats. The process, called catalytic hydrogenation, allows these new products to compete with butter and lard.

Hydrogenation is a process where hydrogen is "bubbled" through liquid oils until the oils become chemically saturated. By adding the extra hydrogen atoms onto the fat molecule, hydrogenation turns an unsaturated fat into a saturated fat, which would be bad enough, but it also produces some pretty strange fat molecules that aren't naturally found in the human food chain.

ILLUSTRATION #1

```
  H H H H H H H H H H H H H H H H   OH
  | | | | | | | | | | | | | | | |   |
H-C-C-C-C-C-C=C-C-C=C-C-C-C-C-C-C-C-C-=O
  | | | | |   | |   | | | | | | |
  H H H H H   H H   H H H H H H H
```

(cis, cis-linoleic acid)

If you'll notice the diagram of a <u>naturally-occurring</u>, fat called linoleic acid, both of the potential binding spots are on the same side.

ILLUSTRATION #2

```
  H H H H H   H H   H H H H H H H H   OH
  | | | | |   | |   | | | | | | | |   |
H-C-C-C-C-C-C=C-C-C=C-C-C-C-C-C-C-C-C=O
  | | | | | |   | | | | | | | | |
  H H H H H H   H H H H H H H H H
```

(trans, trans-linoleic acid)

In the <u>artificially-occurring</u> linoleic acid molecule, created by hydrogenation, the bonding sites are separated-one on either side. It is these man-made fats that have been shown to be a problem. These hydrogenated fats alter the normal production of fatty-like hormones, called prostaglandins.

Prostaglandins were recently discovered and very little is known about their functions. Over one hundred different varieties are known to exist. Preliminary studies indicate they have links to blood pressure, free radical scavenging, transmission of nerve impulses, inflammatory reactions, blood clotting and even cancer.

These unusual fat molecules also change the melting points of the substance. While unrefined, unsaturated fats melt at around 55 degrees F. or less, these beauties won't melt until around 112 degrees F. The fact that they don't smoke or burn at higher temperatures, makes them ideal to use in deep frying (ideal for the owner of the fast food franchise and the heart surgeon that is). These hydrogenated oils have other amazing characteristics as well. You can fry chicken, fish and onion rings all in the same grease and it won't absorb any of the different flavors. And best of all, the customer can't taste any difference when the oil becomes rancid. This last little feature makes it a popular ingredient for cookies and crackers that need a longer shelf-life.

If you want to see just how popular hydrogenated oil has become, start reading the labels of the processed food in your pantry or grocery store. Estimates are that 50% to 75% of fats now consumed in the United States are in hydrogenated form. About 80% of that is hydrogenated soy bean oil.

Hydrogenation is here to stay. Plans are now being made to hydrogenate Omega-3 fish oils and offer it to the public as cholesterol-free margarine and shortening. If and when you do see it in the market, just remember, hydrogenated fat from any source is dangerous to your health!

Speaking of poly-unsaturated fish oils, like Omega-3, generally they can be quite helpful. Fish oils, at the time of this writing, are extracted by pressure, which leaves many of the antioxidant nutrients it contains intact. By encapsulating it quickly, its exposure to oxygen is minimized and the chances of rancidity are lessened. By following a few suggestions you can reduce some of the problems, while gaining the maximum benefits associated with poly-unsaturates.

*Limit the total fats in your diet to no more than 30% (20%-25% would even be better). Break this down further to include 5% poly-unsaturated fats, 15% mono-unsaturated fats and 10% saturated fats. Cut down on your fat. Remember, fats are fats are fats are fats.....regardless of what kind.

*To avoid the problems I've mentioned, get your fats from natural products. The closer to nature, the less processed, the better. Nuts, grains, fruits, vegetables and fish are the best sources.

*Avoid frying or fried foods. Remember, that even the best oils become saturated when heated. If you do end up frying something, don't heat the oil hot enough to make it smoke and always discard it after use. Don't reuse cooking oils. Also, use mono-unsaturated, unhydrogenated oils for frying, such as olive oil.

*Use poly-unsaturated cooking sprays. These sprays generally remain safer than bottled oils.

*Some authorities recommend adding anti-oxidants like the preservative BHT (1 teaspoon per quart of freshly opened oil) or vitamin E (1000 units into each pint).

*Store unused oil in closed containers, preferably in a dark, cool place, such as the refrigerator.

*Add minced or chopped onion or garlic to the oil before cooking to gain the benefit of their anti-oxidant properties.

*Avoid lard, animal fats, shortening and margarine. If you must, use a very small amount of butter, but not margarine. (For a more detailed explanation see the forthcoming section on butter vs. margarine.)

*Try to follow the first two suggestions and remember - a fat, is a fat, is a fat......

There are two other fats we've briefly talked about that could use a little more explanation - mono-saturated fats and saturated fats.

MONO-UNSATURATED OILS

Until recently, these were considered to be more or less neutral, falling somewhere between poly-unsaturates and saturates. New research findings suggest that they can lower L.D.L. cholesterol while leaving H.D.L. levels alone. (Excess of poly-unsaturates and saturates may lower beneficial H.D.L. levels.)

Mono-saturates are somewhat less reactive than poly-unsaturates, however; rancidity and subsequent free radical damage must still be a concern. The best oil in this category is thought to be "virgin" olive oil. With its minimal amount of processing, it still possesses a high degree of protection against oxidation. The same precautionary measures outlined for polyunsaturates should be followed for mono-saturates.

SATURATED FATS

There is little doubt that food producers have used the cholesterol scare as a soap-box to sell their saturated fat products. Two of the cholesterol-free plant oils which are almost totally saturated, now appear in everything from non-dairy coffee creamers to fried corn chips. Coconut and palm oil are the two fats I call "jungle grease". Being saturated, they take longer to oxidize and break down, which increases the shelf life of a product enormously. For this reason, they are common ingredients in foods like cookies and crackers.

Saturated fats occur naturally in animal products and some plant oils like coconut and palm. They are also produced during the process of hydrogenation as previously discussed. Heating of oils is another method that produces saturation. You may remember from high school chemistry, that you can speed up a chemical reaction by increasing the temperature. Deep fat frying provides the ideal temperatures to convert even the most unsaturated oils into saturated fats.

The following chart will help give you some idea of the various types of different fats most oils contain.

OIL	%SATURATED	%UNSATURATED Mono	Poly
Coconut	92	6	2
Palm	53	38	9
Cottonseed	27	18	55
Soybean	15	25	60
Olive	14	77	9
Sesame	14	42	44
Corn	13	25	62
Sunflower	11	21	68
Safflower	9	13	78
Rapeseed (Canadian)	7		

BUTTER VS. MARGARINE

Butter is definitely a highly saturated animal-type fat. Because of this and its higher cost, most people have switched to margarine. The use of small amounts of butter is far safer than chemically hardened oils of margarine. Because of hydrogenation, when you eat margarine, you're eating saturated fats. As discussed fully under the section on hydrogenation, these saturated fats are not compatible with our body chemistry. Animal and human studies done as early as the 1950's, have shown that these man-made hardened oils are dangerous. They contribute to heart and artery disease, cancer, arthritis, nerve disease, and cataracts. They have a higher melting point than our body temperature which allows them to circulate in the bloodstream as a solid fat rather than an oil.

Although it's true, margarine may contain no cholesterol, it has not been shown to lower the risk of heart disease, in fact, it causes the very diseases it is supposed to prevent.

Switch to butter and use it only in moderation. If it seems too expensive, you're definitely using too much! Remember, all fats need to be reduced and a fat, is a fat, is a fat.

MEATS

If you haven't lived under a rock for the last 5 years, you know that the lighter meats like fish, chicken and turkey are lower in both cholesterol and fats, than red meats. This only holds true if you remove the skin and cook poultry by boiling, broiling or baking etc.

FISH

Fish by far seems to be the best choice due to its Omega-3 oil content. This is another incidence where the health benefits of a food can't be judged totally by its cholesterol content. Many of the same types of fish and seafood that lower cholesterol levels, contain high levels of cholesterol themselves. The following chart will help give you some idea of the various amounts of Omega-3 oils in different types of fish.

OMEGA-3 OIL CHART*

Less Than 500mg	Between 500mg-1 gram	Over 1 gram
Atlantic Cod	Carp	Albcore Tuna
Atlantic Pollock	Channel Catfish	American Eel
Brook Trout	Chum Salmon	Anchovy
Haddock	Pacific Halibut	Atlantic Halibut
Northern Pike	Pacific Mackerel	Atlantic Herring
Ocean Perch	Pacific Whiting	Atlantic Salmon
Pacific Cod	Rainbow Trout	Coho Salmon
Rockfish	Red Snapper	King Salmon
Silver Hake	Skipjack Tuna	Lake Trout
Sole	Spot	Pacific Herring
Striped Mullet	Striped Bass	Pink Salmon
Sturgeon	Swordfish	Sablefish
Walleye	Thread Herring	Sardine
Yellowtail	Turbot	Sockeye Salmon
Yellow Perch	Weakfish	Spiny Dogfish
	Wolf fish	

*The total EPA and DHA in approximately a 3-1/2 oz. serving.

PORK

Don't be mislead by the up-coming publicity by the pork industry about the so-called "other white meat". Pork may be white but it's not low fat.

RED MEAT

The beef industry is working on breeding lower fat animals and so far the results are promising. The ultimate success or failure of this venture will as usual, be decided in the market place. In the meantime, if you eat red meat, buy the leaner cuts. Trimming the visible fat before cooking can lower the saturated fats, but it won't do much to lessen the cholesterol. If you remember, cholesterol is in every cell, so removing the fat from any kind of meat will have only a minimal effect on the cholesterol content. For a healthier heart, it's best to cut back on red meat consumption.

Some red meat from animals in the wild, is now being sold in various markets across the country. These meats contain practically no fat and very little cholesterol, however; they are usually more expensive and harder to obtain. With time and new breeding techniques, it's possible these problems will be overcome in the future.

CHAPTER 7

Natural Proven Methods To Lower Cholesterol

Each of the methods outlined in this chapter, have been shown in various clinical and research studies to lower blood cholesterol levels. I don't want to present the idea that cholesterol is the main culprit or cause of heart and artery disease, cholesterol is only one of the factors involved. If anything, it should be considered a diagnostic sign of some deeper underlying problem, such as an improper diet consisting of too many saturated fats or refined carbohydrates (sugar), impaired gallbladder or liver function leading to improper fat digestion, etc. This is one of the reasons competent professional help is advised when dealing with high cholesterol levels, heart disease and circulation problems.

I have no problem recommending or outlining these various techniques for lowering cholesterol levels because most are sound methods of promoting good health in general. Their cholesterol-lowering abilities are generally just welcome side-effects.

A. **FIBER** has the ability to lower cholesterol levels in a rather round about way. It increases the amount of cholesterol that is excreted in daily bowel movements, as well as reduces the amount secreted by the liver. Since fiber is such a popular topic these days, let's take a closer look to see exactly what it does.

Much of the cholesterol produced by the liver is converted into bile acids. The bile acids are stored in the gallbladder to later be dumped into the small intestine to help digest fats. Ultimately they end up in the colon where they are either destroyed or excreted in bowel movements.

Someone on a <u>low</u> fiber diet harbors millions of anti-bile bacteria in their colon. These bacteria attack the bile acids and break them down into several substances, which include, cancer causing chemicals and a toxic product called "Lithocholate". Lithocholate causes the liver to cut back on converting cholesterol to bile acids.

Unfortunately, this causes a couple of serious problems. When less cholesterol is converted to bile acids, it begins to accumulate in the blood stream. Also, less cholesterol reaches the colon where it can be excreted in bowel movements. This is especially dangerous since bowel movements are the body's main method of ridding the body of unwanted cholesterol.

On the other hand, a <u>high</u> fiber diet alters the type of bacteria in the colon. High fiber diets promote friendly bacteria that leave passing bile acids intact. As a result, your body excretes more cholesterol and in effect is "tricked" into wasting more cholesterol by converting it into bile acids. A high fiber diet can lower blood cholesterol levels, protect against colon cancer and even improve fat digestion by providing more bile acids.

In the industrialized nations, the Japanese consume the highest amount of dietary fiber and also have the lowest rate of heart disease deaths. Americans consume the least amount of fiber and have the highest heart disease rate in the world. (Atherosclerosis, May/June, 1982)

VARIOUS TYPES OF FIBER

1. <u>Oats and oat bran</u> seem to be the best fibers for reducing cholesterol. Researchers at the University of Kentucky and the V.A. Medical Center found that oat bran reduced L.D.L. levels over 20 percent in only 11 days. Adding oat bran to the diet each day of those in the study was the only change made to their eating habits or lifestyle. (American Journal of Clinical Nutrition, December 1984)

2. <u>Legumes</u>, such as chick peas or brown beans, can also lower cholesterol levels. The same study mentioned above, showed that substituting brown beans or chick peas for the oat bran produced the same results. Those who stayed on the diet for over two years showed a 29 percent decrease in L.D.L. levels.

3. <u>Apple pectin</u> taken with vitamin C can reduce cholesterol levels in both the liver and the blood. A study performed at the Institute of Human Nutrition in Bratislava,

Czechoslovakia showed that a daily dosage of 450 mg. of vitamin C with 15 grams of citrus pectin lowered L.D.L. cholesterol levels in a period of just six weeks.

4. <u>Grapefruit pectin</u> was shown by a University of Florida study to lower L.D.L. cholesterol levels. Volunteers took a 5 gram capsule three times daily without making any other dietary changes. After eight weeks, total cholesterol dropped 7.6%, L.D.L. cholesterol dropped 10.8% and the L.D.L./H.D.L. ratio 9.8%.

5. <u>Psyllium</u>, taken in the form of Metamucil, started to drop cholesterol in just two weeks. After eight weeks, total cholesterol was down 15%, L.D.L. cholesterol levels dropped 20% and the L.D.L./H.D.L. ratio decreased from 3.2 to 2.6. Patients took one 3.4 gram packet of sugar-free Metamucil three times daily. The results of this study were recently reported to the annual meeting of the Federation of American Societies for Experimental Biology.

6. <u>Guar gum</u> made from the guar plant, which grows in the drier parts of Texas, Mexico, Pakistan and India, has shown to lower cholesterol levels by as much as 25%. Several products that incorporate guar gum are now available in health food stores. Two products: Bio-Guar, a capsule, and Thera-Guar, a drink, are currently marketed by National Bio-Source, Inc.

B. <u>OMEGA-3 OIL</u>, an essential fatty acid, found in fish as well as some plants, vegetables and nuts, has recently become a major tool in helping lower blood cholesterol levels.

1.. <u>Fish and fish oils</u> decreased blood cholesterol 27% to 45% in a study at the Oregon Health Sciences University in Portland, Oregon. Triglyceride levels dropped 64% to 79% in the same study. (JAMA, February 12, 1982)

A study from the Netherlands which followed the eating habits of 872 men for twenty years, concluded that eight ounces of fish per week could cut the risk of heart disease in half.

OMEGA-3 OIL CHART*

Less Than 500mg	Between 500mg-1 gram	Over 1 gram
Atlantic Cod	Carp	Albcore Tuna
Atlantic Pollock	Channel Catfish	American Eel
Brook Trout	Chum Salmon	Anchovy
Haddock	Pacific Halibut	Atlantic Halibut
Northern Pike	Pacific Mackerel	Atlantic Herring
Ocean Perch	Pacific Whiting	Atlantic Salmon
Pacific Cod	Rainbow Trout	Coho Salmon
Rockfish	Red Snapper	King Salmon
Silver Hake	Skipjack Tuna	Lake Trout
Sole	Spot	Pacific Herring
Striped Mullet	Striped Bass	Pink Salmon
Sturgeon	Swordfish	Sablefish
Walleye	Thread Herring	Sardine
Yellowtail	Turbot	Sockeye Salmon
Yellow Perch	Weakfish	Spiny Dogfish
	Wolf fish	

*The total EPA and DHA in approximately a 3-1/2 oz. serving.

2. <u>Flax seed</u>, when ground and combined with vitamin B6 and zinc, can be an excellent source of Omega-3 oil. One heaping teaspoon of ground flax contains about 4 grams of Omega-3. It does require zinc, about 15 mg. a day to be properly metabolized. Flax also contains an anti-B6 factor, and 100 parts per million of B-6 should be added to the ground seed to overcome this problem.

3. <u>Additional food sources</u> of Omega-3 that can help lower cholesterol levels include: the vegetable purslane, spinach and mustard greens, walnuts and walnut oil, wheat germ oil, rapeseed oil (canola), soybean lecithin, tofu, beans, buttermilk and seaweed.

C. <u>ELIMINATING SUGAR</u> from the diet can help lower cholesterol and triglycerides, according to studies at the United States Department of Agriculture Human Nutrition Center in California. This was especially true for those men who had difficulty digesting carbohydrates.

Dr. Yudkin, of the University of London, reported in Lancet, that there was a definite relationship between heart disease and sugar. He found that invariably all of the cholesterol patients he examined had a high intake of carbohydrates, but not necessarily a high fat intake.

Sugar is quickly converted to fat, which raises the level of blood fats, like triglycerides. It also disrupts the production of beneficial intestinal bacteria, which can indirectly increase blood cholesterol levels as explained under the section on fiber.

Sugar also contributes to atherosclerosis. Eating sugar causes the release of insulin, which is required to break it down for energy. However, while the sugar is metabolized quickly, the excess insulin circulates in the bloodstream for hours. This excess insulin is now thought to be one of the factors causing cholesterol and other fats to be deposited along arterial walls.

D. VITAMIN B3 (NIACIN) is one of the most effective natural methods of lowering cholesterol levels. It may also be one of the least expensive. A year's supply of generic niacin retails for less than $50.

Numerous studies have proven its effectiveness. One study details how 3 grams of niacin a day reduced cholesterol levels 26 percent in just two weeks. (J. Angiology, January, 1973)

Some authorities have recommended dosages as high as 7 to 10 grams daily, but niacin caused a flushing and itching sensation of the skin. This effect can be lessened by taking the vitamin only on a full stomach and in small incremental dosages until one's tolerance is built up. (American Journal of Clinical Nutrition 8:480:60)

E. VITAMIN C has proved on numerous occasions to have the ability to drop cholesterol levels. As little as 500 mg. to 1000 mg. daily was reported to significantly lower L.D.L. cholesterol levels. (Lancet ii, 1280-1281, 1971)

Low levels of vitamin C have been linked to plaquing in the arteries, even in the absence of high cholesterol blood levels. What's even more interesting, is that certain animal studies have shown that prolonged vitamin C therapy may even reverse atherosclerotic lesions. (American Journal of Clinical Nutrition, 23:27,1970)

Additional cholesterol lowering effects of vitamin C are reported in the Lancet 2:907:84 and the Yearbook of Nutritional Medicine 1984-5.

F. CALCIUM CARBONATE, in the amount of two grams daily, can reduce both cholesterol and triglyceride levels. Studies performed at both St. Vincent's Hospital in Montclair, New Jersey and the USDA's Human Nutrition Laboratory in Grand Fork, North Dakota have shown that even daily dosages of 1 gram can lower L.D.L. levels.

G. LIGHT EXERCISE in the form of walking, can lower cholesterol. A study involving 30 men walking as little as twenty minutes daily showed positive L.D.L. reductions.

H. VITAMIN E is one of the most prescribed nutrients for cardiovascular problems. Reportedly, it can help lower blood cholesterol and strengthen arterial walls. A March 1982 report in the American Journal of Clinical Pathology, indicated that 800 i.u. of vitamin E daily could help raise the beneficial H.D.L. cholesterol.

I. SELENIUM when combined with vitamin E will reduce fat levels in the blood. A recent study conducted at the Meharry Medical College in Nashville, Tennessee, showed when selenium was combined with vitamin E from soybean oil, rice and wheat germ or corn oil, it has the positive effect of lowering L.D.L. cholesterol.

J. ACTIVATED CHARCOAL, taken as a suspension in water three times daily, reduced L.D.L. levels an average of 41%, while at the same time increased H.D.L. levels. Eight grams of powdered activated charcoal were used for each dose in the Finnish study. Charcoal tablets are also available, but it would take a considerable amount to reach the level of 8 grams three times daily. Having a black stool was the only side-effect noticed; however, it is possible that the charcoal could hinder absorption of some of the fat soluble vitamins (A,D,E,F and K).

K. AVOCADOS are usually avoided because of their naturally high fat content. This may change since recent reports show that 85 percent of the avocado fat is the type that can help lower cholesterol. In one study, adding avocados to the diet lowered cholesterol levels over 40%! (Future Youth, Rodale Press, 1987)

L. LECITHIN has long been regarded as an excellent natural tool for reducing blood fats. An Australian study reported that 20 to 30 grams of lecithin daily could dramatically lower L.D.L. cholesterol levels. (Australian and New Zealand Journal of Medicine, 1977)

In 1977, researchers at the University of Washington found that 36 grams of lecithin taken daily for fifteen weeks decreased L.D.L. cholesterol by 7 percent, while increasing H.D.L. cholesterol by almost 4% (Clinical Research, Vol.25, No.2).

Rutgers Medical School in New Jersey also showed positive results using lecithin to lower cholesterol.

A Swedish study in 1974 showed that patients taking only 1.7 grams per day for nine weeks had H.D.L. levels rise an average of 30 percent.

UCLA School of Medicine conducted research which combined a low-fat diet and 48 grams of lecithin daily. Over a two year period, the cholesterol levels of those involved dropped an average of 22 percent. (American Journal of Surgery, Vol.140, No.3, 1980)

M. **CHROMIUM**, a trace mineral, has been shown to perform amazing feats when it comes to lowering L.D.L. cholesterol levels and improving the total cholesterol/H.D.L. ratio that was discussed earlier.

Chromium, in the form of brewer's yeast (about 2 1/2 teaspoons daily), was given to patients each day for eight weeks. This works out to about 50 micrograms of chromium per day. At the end of the study, serum cholesterol levels dropped from over 240 mg., to under 220 mg., while H.D.L. levels rose.

The starting total cholesterol/H.D.L. ratio averaged 5 and after the study, the ratio fell to 3.9! (Remember that anything above 4.5 is not acceptable for good cardiac health.) As a follow-up, the dosages of brewer's yeast were cut in half and the ratio still declined in almost 80 percent of the patients.

The study was conducted under the direction of J.C. Elwood, Ph.D., at the State University of New York, Upstate Medical Center at Syracuse, New York.

NOTE: Brewer's yeast is considered to be one of the best sources of chromium. The chromium it contains is bound to the Glucose Tolerance Factor (GTF) making it easier for the body to use. If someone has difficulty taking brewer's yeast, it has been suggested that 100 micrograms of chromium in tablet form could probably achieve similar results.

N. **ONIONS** cooked or raw, eaten after a fatty meal, have been shown to prevent an increase in blood cholesterol levels. Their ability to do this isn't fully understood; however, it is known that they don't interfere with fat absorption like many cholesterol lowering agents.

It may be a little too soon to start serving onions for dessert, but considering the growing trend to lower cholesterol it might just come to pass. (The Indian Journal of Nutritional Diet 12:288:75)

O. **GARLIC** can not only help dissolve blood clots following heart attacks, but it also lowers the harmful L.D.L. cholesterol levels. You have to look pretty hard to find a condition that garlic won't help.

P. **SPIRULINA**, the microalgae plankton has reportedly been able to lower abnormally high cholesterol levels. Dr. Tadaya Takeuchi, of the National Medical and Dental University of Tokyo, reported that both, overall liver function and cholesterol levels were significantly improved when twenty-one tablets were taken daily for two to six weeks.

Q. PANGAMIC ACID, also known as B-15, can lower serum cholesterol. (Vitamin B-15: Properties, Functions and Uses; McNaughton Foundation, 1966)

R. BLACK CURRENTS contain both, vitamin C and pectin. Dr. Ginter, with the Institute of Human Nutrition in Bratislava, Czechoslovakia, showed that a combination of these two substances lowered L.D.L. levels after only six weeks. Other foods with this helpful combination include: raspberries, blackberries, tomatoes, strawberries and citrus fruits.

S. COFFEE and other caffeine-containing drinks when eliminated from the diet help lower L.D.L. levels. Decaffeinated coffee doesn't seem to effect cholesterol levels. (University of California, San Diego)

T. CHLORINATED DRINKING WATER has been linked to higher cholesterol levels. Scientists at the Oak Ridge National Laboratory in Anaheim, California, found that of the 2000 women studied, those who used chlorinated drinking water had higher L.D.L. cholesterol levels. (Medical World News 10:13:86)

U. **YUCCA POWDER** can lower both blood pressure and serum cholesterol. Dr.'s Robert Bingham and Bernard Bellow were studying the possible effects this desert plant might have on arthritis, when they stumbled onto these two beneficial side-effects. They reported in the Journal of Applied Nutrition, that natural steroids called caponins, contained by the plant, work similar to a soap to help remove fats from the blood. Yucca tablets, powders, and capsules can now be found in many health food stores.

V. **OLIVE OIL**, one of the mono-saturates, may be even more helpful in lowering L.D.L. cholesterol than safflower and corn oil. (Dr. Scott Grundy at the University of Health Science Center, Dallas, Texas)

W. **ALFALFA** tablets or the sprouted seeds lower L.D.L. Blood levels. This inexpensive product is an excellent source of trace minerals and fiber as well.

X. **BETA-SITOSTEROL** and **ORGANIC GERMANIUM** are two nutritional products that reportedly have been shown to lower blood cholesterol levels. Long term, large scale studies are not available at this time however.

Y. **VITAMIN A** was given along with vitamin D to a group of patients for a period of ten years. At the end of this period, those taking the vitamins had about one-third less cholesterol than untreated patients. The general daily recommendation for vitamin A is around 25,000 i.u. and 400 i.u. for vitamin D. (Med. J. Austrl.,48, 1961)

Z. ZINC helps lower cholesterol and has been used successfully in the treatment of atherosclerosis. It is part of at least 25 substances involved in the digestion and metabolism of fats. (Nutrition Almanac, Nutrion Search, Inc.)

Vandium is another of the trace minerals that inhibits cholesterol production and helps prevent gallstones. (Fed. Proc., 33, 1974, 1773)

CHAPTER 8

Some Not-So-Natural But Interesting Ways To Lower Cholesterol

While doing research for this book, I came across several unusual procedures that just happen to lower cholesterol levels. The two I'll mention here were discovered in the never-ending search to increase our life span.

GH3

In 1950, a drug called procaine was synthesized by Alfred Einhorn. It was developed as a substitute for cocaine to be used as a local anesthetic without cocaine's uncontrollable side effects. In the 1940's it was shown to relieve some forms of arthritis and Dr. Ana Aslan began to use it on many of her patients in Romania. In addition to helping relieve their symptoms, the patients reportedly felt younger, were less depressed and had increased vigor and vitality.

Dr. Aslan improved on the drug and renamed it Gerovital H3 or GH3. With continued use, she started to notice many anti-aging effects like lower cholesterol levels, more elasticity of the skin, fewer wrinkles, re-growth of hair, better memory and normalized blood pressure levels. Ever since Dr. Aslan reported these findings to a German symposium in 1956, GH3 has been a hotbed of controversy.

Studies using procaine have never duplicated the effects of GH3, but several studies using the modified procaine GH3 have supposedly continued to demonstrate its anti-aging properties. The drug GH3 has never been approved for sale in this country, but currently it can be purchased from a company operating out of the Bahamas that will ship it anywhere.

PLASMA PHARESIS

Another unusual method to lower cholesterol levels is called plasma pheresis. This procedure involves the removal of two pints of blood from the body. The blood is then centrifuged under refrigeration. The red blood cells are separated and then returned to the body with two pints of a neutral solution. This forces the body to produce new plasma to replace that removed. It is reported by some that cellular activity is increased, cholesterol metabolism is more efficient, serum cholesterol is lowered and those undergoing the process emerge with new found energy and younger looks.

Plasma pharesis is based on the theory that so-called "death hormones" circulate in the blood stream and their gradual accumulation with age helps lead to degeneration and eventually death. Plasma pharesis is supposed to remove some of these hormones as well as cleanse the blood of toxins, insoluble fats and defective proteins that have accumulated.

If the plasma pharesis theory interests you, but there are none of these type facilities in your area, or there isn't enough funds in your bank account, there is still hope. Proponents of the method recommend donating blood once every month. In about 4 months 50 percent of your total blood volume will have been replaced.

** *Please don't think I am promoting or recommending either of these techniques for lower cholesterol. Both were included only because I found them quite interesting and hope you will also.*

CHAPTER 9

Cholesterol Testing Tips

Estimates are that 25 percent of American adults fall into the high risk category for coronary heart disease. As mentioned before, no one has shown that cholesterol causes heart disease, but your blood levels can be a good diagnostic tool to help determine your risk. There is only one method of determining your serum cholesterol levels and that's through a blood test. There's a few tips you may want to keep in mind when you decide to have a test for serum cholesterol.

1. **Request that your blood levels be checked.** Even if you have the routine blood test given with annual physicals, etc., you may have to request that your blood levels of cholesterol be checked. A recent survey by the National Heart, Lung and Blood Institute, found that less than half of the total population has ever been checked and only about 8% of those were told their levels were too high. If estimates are correct that 25% of

adults have high risk levels (240 mg./dl or above) then more testing definitely needs to be done.

2. Make sure your blood sample is taken properly. There are two ways a blood sample can be taken. With the new portable analyzers, only a drop or two of blood is needed. If blood is taken after a finger prick, make sure you are sitting. Fluids "settle" in the body differently depending on whether you're standing, sitting, or lying down. Also, make sure the technician doesn't milk your finger to extract the blood. This squeezing causes the release of extra fluids surrounding body cells, which will dilute the sample, giving an incorrect low cholesterol reading.

If the blood sample is taken from a vein with a needle and syringe, make sure the tourniquet isn't left on over a minute or two and it should always be released before the blood is drawn. If these precautions aren't followed, the sample will be taken from stagnant blood, which could give false readings.

3. Test at the proper time. Most accurate results are usually obtained by testing either before breakfast or at least two hours after a meal.

Spring and Fall are the best times to have the cholesterol test. Tests will be somewhat lower since generally more vegetables and less fat are eaten around these times. Increased exercise may also play a role.

Stress can increase test levels as much as 10%. Avoid testing during stressful periods.

Menstruating women may receive slightly elevated results, especially toward the end of the cycle as estrogen production decreases. More accurate results can be obtained if testing is avoided during this time.

4. Inform your doctor about any medications you're taking. Drugs for birth control, hormones, high blood pressure medications, diuretics and the epilepsy drug dilantin can alter test results.

5. Ask for your H.D.L./L.D.L. ratio and a copy of all your lab results as well. Remember that a total cholesterol reading includes both, the bad type L.D.L. and the beneficial type H.D.L.

6. If the results are high or questionable, re-test. A recent survey by the college of American Pathologists, showed the inconsistency of even the nations top laboratories. They sent identical samples to 5000 laboratories and only half were able to come within 5 percent of the correct values.

CHAPTER 10

SAVING SOMEONE'S LIFE

Using the technique called, closed-chest cardiopulmonary resuscitation (CPR), you can actually give someone the gift of life by using your own breath and bare hands. CPR when properly administered, can keep the body and brain alive for several hours after the heart has stopped beating or the person has stopped breathing. If you aren't trained in CPR I would highly recommend that you contact your local Red Cross or paramedic group. Both will be happy to arrange for a demonstration and/or training session.

Cardiopulmonary Resuscitation (CPR)

WE'RE FIGHTING FOR YOUR LIFE
American Heart Association

Place victim flat on his/her back on a hard surface.

1. If unconscious, open airway.

Head-tilt/chin-lift.

2. If not breathing, begin rescue breathing.

Give 2 full breaths. If airway is blocked, reposition head and try again to give breaths.
If still blocked, perform abdominal thrusts (Heimlich maneuver).

3. Check carotid pulse.

4. If there is no pulse, begin chest compressions.

Depress sternum 1½ to 2 inches.
Perform 15 compressions (rate: 80–100 per minute) to every 2 full breaths.

Continue uninterrupted until advanced life support is available.

CHAPTER 11

SAVING YOUR OWN LIFE

As strange as it sounds, you may be able to give yourself a modified type of CPR if you ever suffer from heart failure. It was previously thought that pressure to the chest during CPR compressed the heart causing it to pump blood. Researchers have found instead, that the compression of the breast bone causes increased pressure in the chest cavity itself. This forces blood out of the chest area to all parts of the body.

You can keep your own blood circulating by vigorous coughing. It has the same effect by increasing pressure in your chest. In fact, it was found that one cough moves as much blood as two CPR compressions!

To use this technique, you have to have a quick memory. After the heart stops, you have only about 15 seconds before you pass out. Vigorous coughing must start immediately to be successful.

Dr. J. Michael Criley of UCLA, reported the

case in which one of his patients used this coughing technique to stay alive on his way to the hospital after his heart went into fibrillation.

CONCLUSION

After shedding some light on the cholesterol controversy, I hope you will be left with the idea, that practically everything you can do to improve your overall health will improve cholesterol levels also. The human body is not simply a conglomeration of organs, nerves and muscles made from proteins and fat. It is a highly sophisticated miracle, interconnected in ways we'll probably never understand. Our overall health or lack of health is not determined by a single variable like cholesterol, but instead by an intricate balance of the thoughts we think, the air we breathe, the water we drink, the exercise we get and the food we eat.

BIBLIOGRAPHY

Barnes, B. O.; Hypothyroidism, The Unsuspected Illness; Crowell; New York; 1976.

Boyd, William; A Textbook of Pathology; Lea and Febiger; Philadelphia; 1979.

Brown, Kovanen, Goldstein, "Regulation of Plasma Cholesterol by Lipoprotein Receptors," *Science* 212:628-635, 1981.

Edward Gruberg & Stephen Raymond, "Beyond Cholesterol," *Atlantic Monthly*, May 1978(popular magazine).

Encyclopedia of Chemical Technology, 8:805;1965.

Ginter, "Cholesterol: Vitamin C Controls Its Transformation to Bile Acids," *Science* 179:702-704, 1973.

Gordon, et al, "High Density Liproprotein as a Protective Factor Against Coronary Heart Disease: The Framingham Study," *Amer. J. Med.* 62:707-708 May 1977.

Green, "The Synthesis of Fat," *Sci. Amer.* offprint 67, Feb. 1960.

Guyton, Authur C.; A Textbook of Medical Physiology; W. B. Saunders Co.; Philadelphia; 1976.

Karvonen, M. J.; Proceedings of the Nutrition Society, 31:355; 1972.

Naimi, S.; Journal of Nutrition, 86:325; 1965.

Nightingale et al, "Effect of Vitamin C and High Cholesterol Diet on Aortic Atherosclerosis in Dutch Belted Rabbits," paper delivered at FASEB, 62nd Annual Meeting, Atlantic City, NJ, April 9-14, 1978.

Pinckney & Pinckney, *The Cholesterol Controversy*, Sherbourne Press, 1973

Robbins, R. C.; Journal Atherosclerosis Research, 7:3; 1967.

Shank, R. E.; Journal American Dietetics Association, 62:611; 1973.

Slater and Alfin-Slater, "Effect of Dietary Cholesterol on Plasma Cholesterol, HDL Cholesterol, and Triglycerides in Human Subjects," *Fed. Proc.* Abstr. 2194, 1 Mar. 1978.

Sperling, G.; Journal of Nutrition, 25:25 1973.

Thompson & Bortz, "Significance of High-Density Lipoprotein Cholesterol," *J. Am. Ger. Soc.* 26(10):440-442,1978.

Tujunami; Circulation Journal, 35:1559; 1971.

Vander, A.J.; Human Physiology; McGraw; New York; 1975.

Williams, R. J.; Nutrition Against Disease; Pitman; New York; 1971.

Yudkin, J.; South African Medical Journal, 47:44; 1973.

INDEX

A

A,vitamin 59
activated charcoal 55
adrenal glands 8
age 15
alcohol 12
alfalfa 59
animal fat 16
anti-cholesterol drugs 19
anti-oxidants 39,40
anti-oxidant nutrients 36
apple pectin 49
arteries,coronary 14
arthritis 42
atherosclerosis 12,53
avocados 55

B

B3 (niacin),vitamin 53
B15,vitamin 58
beans 52
beans,brown 49
beef 44
Beta-sitosterol 59
bile acids 48
bile salts 8
black currents 58
blood cholesterol levels 23
blood pressure,high 14,25
blood sample 66
blood vessels 9
bran,oat 49
brewers yeast 56
butter 16,29,32,33,40,42,43
buttermilk 52

C

C,vitamin 49,50,53,58
calcium carbonate 54
cancer 42
cataracts 42
cell membrane 7
charcoal,activated 55
chart,Omega 3 oil 44,51
chicken 43
chick peas 49
chlorinated drinking water 58
cholesterol,dietary 4
cholesterol-free products 33
cholesterol,H.D.L. 24
cholesterol,L.D.L. 24,25
cholesterol levels,blood 23
cholesterol testing tips 65
choline 30
chromium 56
cocaine 61
coconut oil 33,41,42
coffee 58
Colestid 19
cookies 41
cooking oils 29
cooking sprays 39
corn oil 33,42
coronary arteries 14
cottonseed oil 42
CPR 69
crackers 41
currents,black 58

D

D,vitamin 59

deep fat frying 41
Diabetes mellitus 25
dietary cholesterol 4
drugs,anti-cholesterol 19

E

E,vitamin 39,54
eggs 29,33
egg yolks 30
estrogen 16,67
exercise 54

F

Farmington Study 4
fat,plastic 18
fats 32
fats,animal 16
fats,mono-unsaturated 34,39,40
fats,poly-unsaturated 34,35,39
fats,saturated 12,21,33,34,39
fat soluble vitamins 19
fat,unsaturated 34
fibrillation 72
fiber 21,48
fish 43,51
fish oils 51
flaxseed 52
free radicals 13,36
fried foods 16
frying,deep fat 41

G

gallbladder 47,48
gallstones 50
garlic 40,57
Gemfibrozil 20
germanium,organic 59
Gervitol H3 61,62
GH3 61,62
glands,adrenal 8
grapefruit pectin 50
guar gum 50
Guidelines,National Institute of Health 24

H

heart attack 10,12
H.D.L.(high density lipoproteins) 23
H.D.L. cholesterol 24
hemorrhage 14
heredity 15
high blood pressure 14,25
hormones 15
hydrogenated oil 38
hydrogenation 36

I

inflammation 10
insulin 12,53
intestine,small 4

J

"jungle grease" 41

K

kidney failure 12

L

lard 16,40
L.D.L.(low density lipoproteins) 23
L.D.L.cholesterol 24,25
lecithin 30,55
lecithin,soybean 52
legumes 49
Lithocholate 48
liver 4,8,47,48
Lovastatin 20

M

margarine 16,29,32,40,43
meat,red 45
melting point 38,42
membrane,cell 7
menopause 16
menstruating 67
metamucil 50
milk 33
mono-unsaturated fats 34,39,40
mustard greens 52

N

Nathan Pritikin 17
National Institute of Health Cholesterol Guidelines 24
nerve disease 42
(niacin),vitamin B3 53

O

oat bran 49
oats,49
obesity 25
oil 32
oil,coconut 33,41,42
oil,corn 33,42
oil,cottonseed 42
oil,hydrogenated 38
oil,rapeseed 42,52
oil,safflower 42
oil,sesame 42
oil,soybean 33,42
oil,sunflower 42
oil,palm 33,41
oil,Omega 3 32,35,38,43,50
oil,"Virgin" olive 40
oil,walnut 52
oil,wheat germ 52
oils,cooking 29
oils,fish 51
Olestra 18
olive oil 40,42,59
Omega 3 oil 32,35,38,43,50
Omega 3 oil chart 44,51
onion 40,57
organic germanium 59
ovaries 8

P

palm oil 33,41,42
pangamic acid 58
peas,chick 49
pectin 58
pectin,apple 49
pectin,grapefruit 50
plasma pharesis 62
plastic fat 18
poly-unsaturated fats 34,35,39
pork 44
Pritikin,Nathan 17
procaine 61,62
prostaglandins 37
psyllium 50
purslane 52

Q

Questran 19

R

rapeseed oil 42,52
red meat 45

S

safflower oil 42
salts,bile 8

saturated fats 12,21,33,34,35,42
seaweed 52
seed,flax 52
selenium 54
sesame oil 42
shortening 40
small intestine 4
smoking 12
soybean lecithin 52
soybean oil 33,42
spinach 52
spirulina 57
sprays,cooking 39
steroid 8,59
sterol 3
stress 67
stroke 10
study,Farmington 4
sucrose polyester 18
sugar 52,53
sunflower oil 42

T

testes 8
testing tips,cholesterol 65
tofu 52
triglycerides 25,51,52
turkey 43

U

unsaturated fat 34

V

vandium 60
"virgin" olive oil 40
vitamin A 59
vitamin B3 (niacin) 53
vitamin B15 58
vitamin C 49,50,53,58
vitamin D 59
vitamin E 39,54
vitamins,fat soluble 19

W

walnut oil 52
walnuts 52
water,chlorinated drinking 58
wheat germ oil 52

XYZ

yeast,brewers 56
yucca powder 59
zinc 60

NOTES

NOTES

NOTES